Managing waste through time

The way people have dealt with garbage has changed throughout time. There have been many clever inventions and new ideas for dealing with waste.

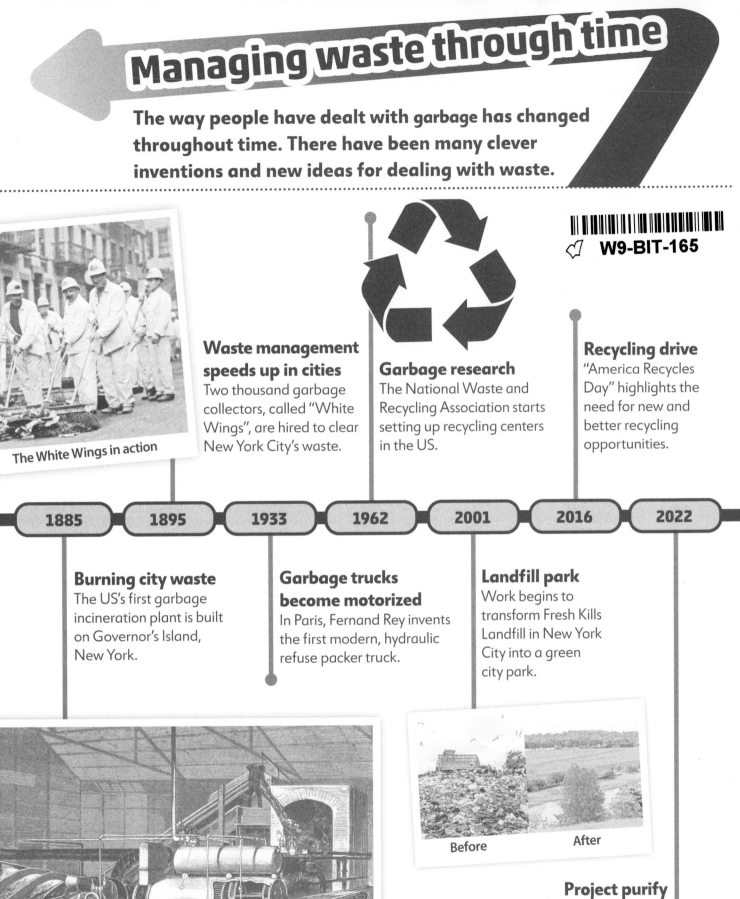

Waste management speeds up in cities
Two thousand garbage collectors, called "White Wings", are hired to clear New York City's waste.

The White Wings in action

Garbage research
The National Waste and Recycling Association starts setting up recycling centers in the US.

Recycling drive
"America Recycles Day" highlights the need for new and better recycling opportunities.

1885 · 1895 · 1933 · 1962 · 2001 · 2016 · 2022

Burning city waste
The US's first garbage incineration plant is built on Governor's Island, New York.

Garbage trucks become motorized
In Paris, Fernand Rey invents the first modern, hydraulic refuse packer truck.

Landfill park
Work begins to transform Fresh Kills Landfill in New York City into a green city park.

Before   After

Project purify
A new initiative in Singapore will be rolled out to collect, treat, and purify waste water.

Things to find out:

DK findout!
Garbage

Author: Anita Ganeri
Consultant: Dr. Stephen Burnley

DK | Penguin Random House

Senior editors Carrie Love, Roohi Sehgal
Project designers Rachael Hare, Jaileen Kaur
Senior art editor Nidhi Mehra
US Senior editor Shannon Beatty
DTP designers Sachin Gupta, Vijay Kandwal
Project picture researcher Sakshi Saluja
Jacket coordinator Issy Walsh
Jacket designer Debangshi Basu
Jacket editor Radhika Haswani
Managing editors Penny Smith, Monica Saigal
Managing art editors Mabel Chan, Romi Chakraborty
Production editor Rob Dunn
Production controller Ena Matagic
Delhi creative heads Glenda Fernandes, Malavika Talukder
Publishing manager Francesca Young
Creative director Helen Senior
Publishing director Sarah Larter
Educational consultant Jacqueline Harris

First American Edition, 2021
Published in the United States by DK Publishing
1450 Broadway, Suite 801, New York, NY 10018

A catalog record for this book
is available from the Library of Congress.
ISBN: 978-0-7440-3698-5 (Hardcover)
ISBN: 978-0-7440-3697-8 (Paperback)

DK books are available at special discounts when purchased in bulk for sales promotions, premiums, fund-raising, or educational use. For details, contact: DK Publishing Special Markets, 1450 Broadway, Suite 801, New York, NY 10018
SpecialSales@dk.com

Printed and bound in China

For the curious
www.dk.com

MIX
Paper from
responsible sources
FSC™ C018179
www.fsc.org

This book was made with Forest Stewardship Council™ certified paper— one small step in DK's commitment to a sustainable future. For more information go to www.dk.com/our-green-pledge

Contents

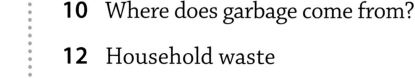
Used coffee grounds

Rotten vegetables

Rubber boots

Dump truck

Craft made from waste

Plastic bag

Plastic waste harms coral reefs.

Antique violin found in trash

Garbage can and bags

What is garbage?

Every day, we throw away millions and millions of tons of garbage. Some of this waste is biodegradable. This means that it rots away and soaks into the soil. But a lot of the waste never breaks down, and it is turning the Earth into a giant trash dump.

Garbage patch
This floating garbage dump in the reservoir by the Carpathian mountain range in Europe shows how plastic waste clumps together easily. It breaks down so slowly in water that it may never disappear entirely.

The problem with waste

Getting rid of so much garbage is a serious problem around the world. When waste is simply thrown away, it can be deadly to animals, plants, and the environment.

A seal trapped in trash in the Pacific Ocean

Animals
Sea animals can get tangled up in garbage that's dumped in the sea, such as old fishing nets and plastic bags. They can also swallow garbage that poisons their bodies.

Blooming algae

Fertilizer from farmers' fields can wash into rivers when it rains. Tiny river plants called algae use it to grow. Algae end up covering the rivers and harming the wildlife.

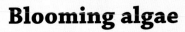

Children playing near trash in Cambodia

Burning waste in Nigeria

Humans
Rotting trash is a breeding ground for diseases that can be deadly to humans. But getting rid of waste safely is expensive, so it can be hard to manage.

Environment
Some garbage is burned, releasing toxic gases into the air. This causes problems for the environment, and can lead to other fires breaking out.

What's hidden in a midden?

Long ago, people threw their everyday waste onto large heaps in their villages or outside their houses. These ancient trash dumps are called middens. Many middens can still be found. By studying the garbage, archaeologists can learn a lot about how people lived.

WOW!

The **oldest** known shell middens are more than **140,000 years old**.

Charcoal and ash
Piles of charcoal and ash are proof that the people who built the midden used fire as part of their daily lives.

Tools
Broken tools give valuable information about what materials people used, what work they did, and how they hunted and prepared food.

Looking at a midden
A midden might just look like a lot of junk, but its contents provide archaeologists with useful clues about how people lived in the past.

Pottery
Pieces of old pottery can be put back together to show the sorts of items that people used for cooking, eating, and storing food and drink.

Estimating age
Scientists can discover how old bones are by using carbon dating. This process is used to figure out the age of a midden.

Bone material is analyzed

Shells
Along the coast, middens are mostly made up of shells from oysters, clams, and mussels. These are the leftovers of meals eaten by hunters.

Skulls and bones
The remains of animals, such as rats, are often found in middens. The garbage was a tempting source of food for wild animals.

Trash or treasure?

Sometimes, amazing and unexpected treasures are found in the trash, from bars of gold to rare antiques. But garbage doesn't always have to be worth a fortune to be valuable. One person's trash could be another person's way of making a living.

Living on landfills

In some places where it's hard to find a job, poor people sometimes make their homes in trash dumps. They risk their health, searching among the piles of waste to find items that they can sell to make money.

People searching a dump site in Thailand

ICONIC HAT

In 2019, a man in the UK was working at a trash dump when he spotted some interesting items. He asked an expert and found out that the items all belonged to Sir Winston Churchill, an important historical figure. The collection included hats, letters, a signed photograph, and even an old cigar.

Winston Churchill was a former Prime Minister of the UK.

GOLD FOUND

In South Korea in 2018, a janitor at an airport found seven gold bars inside a trash can. The janitor wrapped them in newspaper and handed them in to the police. The gold was worth over $300,000!

ANCIENT MAYAN ARTIFACTS

In 2004 in New York, Nick DiMola cleared out the apartment of an artist who had died. He found a cardboard barrel and stored it for a few years. When he finally opened it up, he found almost $20,000 worth of ancient Mayan artifacts dating between 300 BCE and 500 CE.

The items were carefully stored in straw and kept safely in the cardboard barrel.

Baseball cards

In 2016, a family in the United States was clearing out a late relative's house when they found seven old baseball cards. They almost threw them in the garbage, but it's a good thing they didn't. The cards were over 100 years old, and were extremely rare collectibles for baseball card fans. They were worth almost a million dollars!

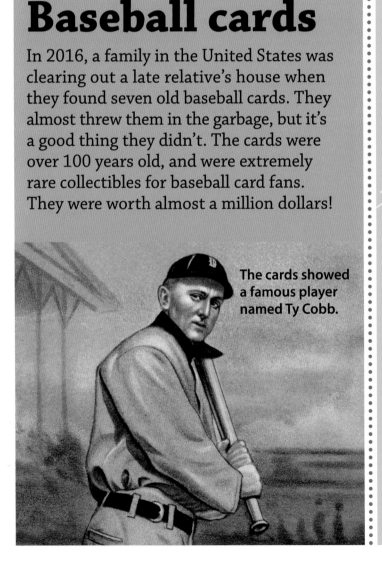

The cards showed a famous player named Ty Cobb.

A similar violin from Pedrazzini's collection.

Antique violin

In Texas, a man spotted an old violin by the side of the road. He discovered that the instrument once belonged to a famous Italian violin-maker, Giuseppe Pedrazzini, and was worth a lot of money!

Where does garbage come from?

Almost everything we do in our daily lives produces garbage of some sort. Other waste comes from factories, industries, farms, and stores, not to mention e-waste—electronic items such as old smartphones, broken computers, and other household appliances.

Wasted carrots and potatoes

Rotten vegetables can't be eaten.

Companies make a lot of commercial waste everyday.

Agricultural waste
This is waste from farms. It includes things such as crops that have gone rotten, animal waste, and poisonous fertilizers and pesticides. These sometimes wash into rivers when it rains.

It's important to get rid of industrial waste safely.

Remember to recycle as much as you can.

Household waste

From leftover food and plastic packaging, to used coffee filters and old clothes, homes around the world throw out tons of waste on a daily basis. It's good to be mindful of how much you throw away.

Categories of waste

Waste can be sorted into two types, depending on what it is made from.

WET

DRY

Wet waste includes things like rotten food.

Dry waste are things like glass and plastic.

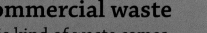

Always dispose of e-waste safely.

Commercial waste

This kind of waste comes from stores, offices, and schools, which produce a lot of trash every day. Their waste is usually made up of cardboard, paper, plastic, and broken equipment. Lots of these companies try to recycle to reduce their waste.

Electronic waste

Electronic waste is also known as "e-waste." It includes things like old smartphones and televisions, used batteries, and unwanted fridges and freezers. These must be disposed of in a special, safe way.

Industrial waste

Factories and industries produce waste such as chemicals, which can be dangerous to people and animals. This waste may be dumped in seas and rivers, which harms the environment.

Household waste

Think of all the garbage you and your family produce at home every single day. What did you throw away today? Maybe a magazine, food, or a used toothbrush. A lot of this waste could be used for something else, or recycled into something new.

Old or unwanted clothes can be mended, donated to charity, or put in a clothes bank to be recycled.

Paper
Waste paper includes egg cartons, wrapping paper, cardboard, and newspapers. Lots of it can be recycled instead of being thrown away. Find out if paper is collected from your home or if you have a recycling center nearby.

Metal
Metal is used to make many household items, such as cutlery, aluminum foil, and tin cans. Instead of throwing them away, you can recycle things made from steel or aluminum.

Plastic

Millions of plastic bottles, bags, and food packaging are thrown away every day. Some types of plastic can be recycled, but lots of plastic is simply dumped. It's always best to recycle if you can.

Why not work with an adult to use the wood from unwanted items in your home to make something new?

Glass

Glass bottles and jars are easy to recycle, but not all towns and cities collect glass waste. They can also be reused at home as containers for spices, rice, and other food and drink. Always wash them before reusing them.

Some parts of household waste are unpleasant or even harmful. Food waste smells, attracts flies, and can contain harmful microbes. Some people also put pet poop and used diapers in their garbage cans.

If your old shoes or boots are still in good condition, sort them into pairs and put them in a shoe recycling bank.

1 Farms

On the farm, crops can be damaged by farm machinery, insects, and diseases. Some of these crops are not safe for people to eat; others could be used, but supermarkets are unwilling to sell them.

2

The stages of food waste

In many countries, people buy much more food than they need. Often this food goes rotten or passes its sell-by date, and ends up being thrown away instead of eaten. A huge amount of food is also wasted as it passes through the different stages that take it from the farm to the dining table.

Transport and storage

Crops can be damaged on the way to the warehouse, or during storage before processing. Animals raised for meat may also suffer during transport.

! **REALLY?**

Every year, about **one-third** of all the food produced by farms is **wasted.**

3 Processing

At the factory, raw fruit, vegetables, and other ingredients are packaged into cans, cartons, and bags. This process creates lots of peelings, pulp, and other waste.

Distribution

Once food reaches stores or markets, it is put on display for sale. If it still hasn't been bought by the time it reaches its sell-by date, it just gets thrown away.

5 Consumption

In many homes, schools, offices, and restaurants, uneaten food that's left on our plates is thrown into the garbage and wasted.

Electronic waste

E-waste includes electronic equipment, such as old computers and mobile phones. It contains harmful metals and chemicals, and needs to be recycled or disposed of carefully. E-waste can include valuable items, such as gold, so it's important to reuse this as it will eventually run out.

! WOW!

In 2019, more than **50 million** tons of **e-waste** was made around the world.

Cooling equipment
Cooling equipment includes fridges and freezers. They contain substances that can be harmful to the environment.

Screens and monitors
Broken screens from TVs and other gadgets can be sent back to the company that made them, or recycled at some stores.

Lamps
Lamps and lighting, such as fluorescent tubes, should never be put into general waste. They need to be disposed of safely by a recycling service.

How can we reduce e-waste?

Instead of recycling or throwing away your broken electronic gadgets, why not repair them or pass them on to someone who might need them? This is a much better way to reduce the amount of e-waste we make.

Broken gadgets can often be repaired and used for longer.

Some stores accept old gadgets and games to sell them for a lower price.

Large equipment
Large equipment includes washing machines and ovens. These need to be taken to a specialist service to be recycled.

Small devices
Small devices can be taken to the local household waste site. These are things like vacuum cleaners, toasters, microwaves, and radios.

Communication devices
Items such as mobile phones and smart watches can be wiped of personal data before being sold on.

Where does waste go?

Not all of your waste ends up in landfills. To protect the environment, more waste is being sorted and sent to recycling centers, composters, and waste-energy plants. There's still a long way to go before we have a zero-waste world, but it's a start.

General waste

This waste cannot be recycled. It is transported by garbage trucks to a landfill site to be buried, or to an incineration plant where it is burned.

Small trucks come to pick up recyclables.

Recyclable waste

In many countries people sort recyclable waste into different kinds, such as glass, plastic, and paper. Each kind of waste is put into a different colored bin.

Organic waste

This is waste that originally comes from plants and animals. It includes rotten food, farm waste, and garden waste such as grass cuttings.

A waste transfer station, where waste is loaded onto trucks for transport to a disposal site

Landfill

Here waste is dumped in a hole and buried. Some landfill sites are huge, more than 150 times the size of a soccer field.

Incineration plant

In these plants, waste is burned at high temperatures. This produces energy that is used to make electricity.

Recycling plant

At a recycling plant, the recyclable waste is sorted even further by type, and packed up ready to be turned into new products.

Gardens and farms

Some organic waste can be turned into compost and used to fertilize gardens and fields. It provides rich nourishment for new plants.

REALLY?

Millions of tons of metal, paper, and plastic are sent to other countries to be recycled.

At home

Flushing the toilet, brushing your teeth, washing fruit in the sink—all of these use water that goes to waste. There are easy ways to save water, such as turning off the faucet when you don't need it. Lots of water would be saved if people did this.

On the farm

All over the world, farms use enormous quantities of water to grow crops, such as the rice in these flooded paddy fields. Water can be saved if farms cultivate plants or grains that require less watering.

Collect water to use in the garden
When you wash fruit and vegetables, collect the wasted water and use it to water your garden instead.

Irrigate using sprinklers
One way of cutting down water waste is to water fields so that the correct amount is used and it is applied close to the plant root systems.

Water waste

Think of all the water you use at home every day. Add to that the water used by farms, factories, and businesses. Water is used to flush toilets, grow crops, and make products, but a huge amount goes to waste in the process. So, what steps can we take to save this very precious resource?

In factories

Factories use water in their manufacturing processes to clean products and to cool down machinery. Millions of gallons of wastewater may be pumped straight into rivers, but could be reused instead of being dumped?

Commercial use

Every day, water is used in hundreds of different ways in stores, laundromats, hotels, restaurants, gyms, and other businesses. At some car washes, cars are hosed down with water, much of which goes to waste.

Treat and reuse wastewater
Wastewater from factories can be treated and reused for flushing toilets and cleaning floors.

Use water-saving equipment
Fitting a sprayer to a hose saves water. The trigger makes the water easier to turn on and off.

! WOW!

The process of **making a cotton T-shirt** uses about **475 gallons (1,800 liters)** of water.

Reusing wastewater in parched Jordan

Some cities have set up plants where wastewater can be treated and turned into safe drinking water. This is a good option for places where water is scarce.

Sewage treatment plant in Madaba, Jordan

Where does poop go?

Everyone needs to go for a poop sometimes. It's a normal process for human beings and animals alike. In nature, animal poop makes the soil rich so that plants can grow. Human poop, however, is different. It has to be disposed of in a safe way.

Poop is flushed

After you poop, you flush the toilet. The poop goes into a large sewage pipe with pee, toilet paper, and other wastewater.

Large waste includes things like diapers and cotton swabs, which should not be flushed.

Bigger waste is separated

The wastewater goes to a sewage plant to be cleaned. It passes through giant sieves to remove any large waste that has been flushed.

Grit is removed

Next, the wastewater is collected in a big basin, called the grit chamber. This is where any small pieces of grit like sand and gravel are removed.

Poop is removed

The wastewater is stored in a large tank. Poop settles on the bottom of the tank since it is heavy.

The poop is pumped away from the water tank in pipes.

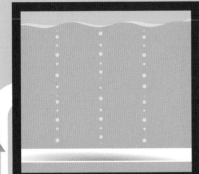

Air treatment
Air is blown into the wastewater, which helps the useful bacteria to break down organic waste and kill harmful bacteria.

Water is cleaned
The wastewater is sprayed over a deep bed of stones. Good bacteria, air, and sunlight break down any remaining waste and kill the final harmful bacteria.

The water is disinfected with chlorine, UV, or ozone.

! WOW!

Dung beetles eat animal poop. They roll it into round balls.

Clean water is returned to rivers
The cleaned water flows back into a nearby river, or to the ocean if you live near the coast.

Sludge treatment
The poop and other solids that settle on the bottom of the tank are called sludge. They are pumped away.

Animal poop
Some poop can spread diseases. It's important to clean up your pet's poop safely. Clean your cat's litter box often. Pick up dog poop using biodegradable poop bags.

Cat poop

Dog poop

Hazardous waste

Hazardous waste is waste that could be harmful to people, animals, or the environment. It can come from factories, hospitals, mines, farms, laboratories, and our homes. Each type of hazardous waste must be disposed of very carefully to make it safe. You can find out about the different types of hazardous waste below.

Waste oil and fuel

Waste oil, gasoline, and other motor fluids contain dangerous chemicals. They should never be put out with your everyday trash, or poured down the drain.

Dispose oil at designated places

Old paint cans should be left to dry out.

Paint wastes

Some decorating products, especially paint, are hazardous. Liquid paint is banned from landfills. You need to let it dry up and go hard, then you can take it to your local disposal site.

Batteries

If used batteries end up in a landfill, the chemicals they contain can leak and cause harm. Put them in a recycling bin. They will be taken apart, and the materials can be used again.

Old batteries leak chemicals

Chemical waste

Many large factories produce chemical waste, which can be harmful to animals, people, and the environment. The factories should always make sure to dispose of their chemical waste safely.

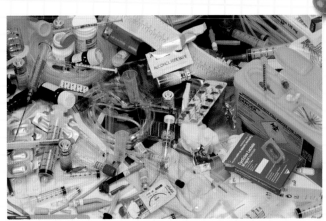
Medical waste goes in special containers.

Chemical waste may harm factory workers.

Medical waste

Hospitals and doctors' offices use medical equipment, such as needles and syringes. These should be put into a special "sharps" container immediately after use.

Nuclear waste

Nuclear waste can remain dangerous for thousands of years. It takes lots of time for it to become safe. Some nuclear waste is stored deep underground, or mixed with concrete and stored in steel drums.

The chemicals meant for farms can run off into rivers.

Nuclear waste is incredibly dangerous.

Pesticides and fertilizers

Some farms spray chemicals on their fields. This helps to grow the crops and kill pests. Excess chemicals can run off into rivers and streams and cause a lot of harm to people and animals.

Space junk

Trash is a big problem on our planet, but we're also turning space into a junkyard. Thousands of pieces of litter orbit the Earth, from old satellites to tiny chips of space paint. Even the smallest pieces of space junk can be dangerous if they crash into working satellites and spacecraft.

Leftover rockets

Abandoned rockets are another major hazard in space. These are rockets that have been fired off in stages, to launch space shuttles and satellites into orbit.

Waste in space

Space junk can zip along at up to 18,000 mph (28,960 kph). At this speed, even a piece the size of a grape could badly damage spacecraft like the International Space Station. Space waste is tracked to help prevent any collisions.

Trash tracking
The Goldstone Observatory in the Mojave Desert, California uses its antenna to help track space junk.

Goldstone's 230-foot (70-m) antenna detects 2 mm debris at altitudes below 621 miles (1,000 km).

Small scraps

Large items of space junk can be tracked. But hundreds of thousands of pieces, such as nuts and bolts, garbage bags, spatulas, screwdrivers, and even gloves, are too small to track.

Stray satellites

Old satellites are left in space when they break down or come to the end of their missions. If two satellites collide, they can smash into thousands of new pieces of space junk.

Low-Earth orbit (LOE)

Most space junk sits in the LOE, within 1,250 miles (2,012 km) of the Earth's surface. This zone is also home to the International Space Station (ISS). A collision could be disastrous.

Cleaning up
In 2016, the H-II Transfer Vehicle, Kounotori was launched by Japan to help clear space debris using a half a mile (0.8 kilometers) long tether.

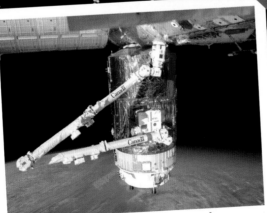

The Kounotori docked on the International Space Station

! WOW!

About **500,000** marble-sized debris objects are believed to be in Earth's orbit!

What to wear

What's in your wardrobe? Do you have lots of clothes that you never wear? Making clothes uses up energy, water, and materials, and releases harmful gases into the atmosphere. We buy and throw away millions of clothing items each year. Many could be fixed, reused, or recycled instead.

Polyester

Polyester is an artificial material used in many clothes. It is made from tiny plastic fibers. These can travel through the sewage system, and end up polluting rivers and oceans.

REALLY?

Every minute, an average of 60 garbage trucks full of clothes are dumped in a landfill, or burned.

Chemical colors

The chemical dyes used to color clothes are one of the main causes of water pollution.

Denim

Denim is a tough cotton material used for jeans, jackets, and dungarees. Producing a pair of jeans uses around 660 gallons (3,000 liters) of water, as well as dye and energy.

Cotton

Cotton is a natural fabric made from the cotton plant. It takes huge amounts of water and pesticides to grow. Some fashion companies use cotton that has been grown sustainably.

Fur

Fake fur has saved the lives of thousands of animals, but it takes hundreds of years to rot away. It can also shed plastic particles and chemicals that cause water pollution.

Leather

Leather is made from animal skin. Raising cattle for leather results in huge amounts of methane, one of the greenhouse gases that is causing global warming.

The problem of plastic

Our lives are full of plastic, from bags and bottles to pots and packaging. Plastic is a useful material, but we need to be responsible about how we use it. Most waste plastic ends up in landfills, where it never rots away. So, what can we do? Find out about the types of plastic and the challenges of recycling them.

Polyethylene Terephthalate (PET)

PET is one of the most common types of plastic. It is often used to make bottles. PET can be widely recycled. The recycled plastic is used to make things like carpets and winter fleeces. Polyethylene is another plastic that we use in lots of packaging material. Like PET, this material can be recycled into new plastic goods.

! WOW!

Around **80 percent** of all the plastic ever made **still exists** today.

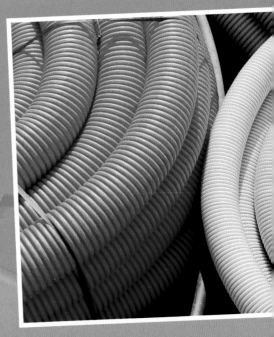

Polystyrene (PS)

Lots of drinking cups, egg cartons, and food packaging are made from the lightweight plastic called PS. It is difficult to recycle, and easily crumbles into tiny, harmful pieces.

Lots of restaurants are trying to cut down their use of polystyrene.

Polypropylene (PP)

A very useful plastic, PP is hard-wearing and waterproof. It's used for items such as baskets, colanders, car parts, outdoor rugs, and cutting boards. It also has a high melting point, so it's used to make handles for saucepans. Many recycling centers take PP, but check first.

Recycling centers have different rules about plastic caps.

Polyvinyl chloride (PVC)

PVC is found in food wrapping, credit cards, medical equipment, and plumbing pipes. It contains harmful chemicals and is difficult to recycle, so we need to cut down our use of PVC.

Save the seas

Covering two-thirds of the Earth, the oceans have been turned into the world's biggest garbage dump. From oil spilled from tankers and chemicals leaking from land to human sewage and waste plastic, millions of tons of trash are dumped in the seas every year, with deadly results.

Danger to marine life
Sea animals mistake plastic waste for food, and can be poisoned when it breaks down in their stomachs. They can also get tangled up in old fishing gear.

Human health
A beach covered in rotting waste and litter is a health hazard to humans and animals. In places where people earn their living from tourism, the sight can stop tourists from visiting.

Habitats in danger
Fragile habitats, such as coral reefs, are being destroyed by plastic litter and oil spills. Corals need clean water to grow. In dirty water, they can get more diseases.

Great Pacific
Garbage Patch

Garbage patch

A giant raft of garbage lies in the North Pacific Ocean. Known as the Great Pacific Garbage Patch, it is made up of around 1.8 trillion pieces of plastic, washed there on ocean currents.

How can we help?

You can help keep the seas healthy by taking part in a beach cleanup with your family. Remember to wear rubber gloves, and look out for glass and other sharp objects. Follow all the instructions from the leader.

! REALLY?

In Mumbai, **India**, about 22 million lb (10 million kg) of **garbage** was **cleared** from one beach.

Large landfills

A landfill site is a huge dump, where garbage is piled up or dumped into a giant hole in the ground. As its name suggests, it's where garbage fills the land. In the past, places like old quarries were used for landfills. We are now running out of suitable spaces to make into new landfills.

One of the world's largest dumps

Bantar Gebang is a massive landfill site in Indonesia. Every day, 6,613 tons (6,000 metric tons) of garbage is dumped there. Some people live on the site, picking through the trash to find things to sell. These people are usually very poor and suffer from many diseases and injuries from working on the landfill.

Disadvantages of landfills

Landfills are a practical way of getting rid of waste. Many are well managed and have strict rules about what they can take. But there are also downsides and dangers.

Danger for human beings
For people living near landfill sites, they can be a health hazard. Rats and harmful bacteria are likely to breed there, and can spread deadly diseases to humans.

Threat to animals
To birds and other animals, landfill sites can look like a good source of food. However, animals can easily be poisoned by toxic materials or pieces of plastic.

Water contamination
As the garbage in the landfill rots, it produces toxic liquids, called leachate. If these leak through the soil, they can pollute groundwater and nearby streams.

Dangerous gases
As the garbage rots, many gases are released. Some have an unpleasant sweet, sickly smell whereas others, such as methane and carbon dioxide, are greenhouse gases. Methane can cause explosions.

Using up Earth's resources

A natural resource is something we use that is found in nature. Among the Earth's natural resources are water, fossil fuels, wood, and minerals. Some resources take millions of years to form. Today, we are using them up faster than they can be renewed.

Water

Water is vital for humans. We use it for drinking, washing, cooking, cleaning, and watering crops. But there is only a limited amount of fresh water. We need to stop wasting precious supplies and think about how we use water.

Fossil fuels

Fossil fuels include coal, oil, and natural gas. They formed from tiny plants and animals millions of years ago. We burn huge amounts of fossil fuels every day, in cars, homes, and factories.

Recovering resources from waste

The Earth's resources are running out fast. We need to cut down on how much we use. By recycling waste, such as paper, we can reduce the need for raw materials, and for energy.

Plastic bottles, tin cans, and cardboard are some other things we can recycle.

Wood

More than half of the land was once covered in forests. Today, less than half of this is left. All over the world, forests are being cut down and cleared, to harvest wood, gather other resources, and to provide land for farming.

Minerals

Minerals are found in rocks, often deep underground. They include valuable gemstones, such as diamonds, and precious metals, such as gold. Mineral mining can damage the environment.

Problem of pollution

Pollution is when the environment is damaged or made dirty by something harmful. Pollution can happen in water, on land, and in air. It is dangerous to people and animals. It comes from factories and other activities, such as driving a car, or littering.

Different materials in dumps contain toxins.

Land pollution

Chemicals may seep into the soil at landfill sites. Poisonous chemicals used in factories and on farms can also build up in the ground. The rain washes them from the soil into rivers and the sea.

Water pollution

Rivers, lakes, ponds, and oceans can also be polluted by sewage, garbage, chemicals, and oil. Oil, spilled from pipes or tankers, is a major problem at sea. It coats and kills seabirds and animals.

Polluted water has ruined this bird's habitat.

Air pollution

Most air pollution comes from the burning of fossil fuels in cars, power stations, and factories. This releases harmful gases, ash, and soot into the air, causing breathing problems for some people.

Smoke can contain many different harmful chemicals.

Aircraft engines cause noise pollution.

Noise pollution

You can't see noise pollution, but it can still be harmful. Very loud sounds at night can make it difficult to sleep, and can make you feel anxious. At sea, the noise from ships can disrupt animal behavior.

Light pollution

The bright lights coming from cities at night can disturb animals and humans alike. The light can confuse migrating birds flying over cities, and baby turtles hatching on beaches.

One impact of light pollution is that it makes it harder to see stars in the night sky.

The greenhouse gas gang

The atmosphere is a layer of gases that surrounds the Earth. Some of these gases trap the sun's heat, keeping the Earth warm. Human activities are adding more of these gases, causing the Earth's temperature to rise.

Heat energy from the sun's rays

Energy reflected back by the Earth's atmosphere

Energy reflected back by the clouds

Earth's atmosphere

Heat from the sun
Heat and light from the sun travel through space to the Earth. The sun makes it possible for life to exist.

Some energy is reflected back by the Earth.

Some of the sun's energy is absorbed by oceans and land.

Global warming

An increase in greenhouse gases is making the Earth warmer. More heat is being trapped by the gases. This is leading to changes in the climate across the world, and to more extreme weather.

Flooded coastlines in Bangladesh

Melting ice in Greenland

Greenhouse gases

Greenhouse gases trap heat, just like the glass in a greenhouse. They keep the Earth at just the right temperature. These "greenhouse" gases include carbon dioxide, methane, and ozone.

Reflected energy
Some of the sun's energy is absorbed. However, a lot of it is reflected back into the atmosphere as heat.

Energy reflected into space

Some of the energy is trapped by greenhouse gases. It heats up and goes back to the Earth.

Greenhouse gases trap some of the energy that is reflected back by the Earth.

! WOW!

If the Earth warms up by just 35°F (1.5°C), around one-third of species could **die out**.

41

Renewable energy

Most of the energy we use to heat and power our homes, schools, and factories comes from fossil fuels (coal, oil, and gas). These fuels are running out fast, and people are looking at other sources of energy that are renewable, which means that they can be made again and again.

Wind energy

Wind turbines, like these in California, capture the energy of the wind, and convert it into electricity. A large group of wind turbines is known as a wind farm.

Hydro energy

A dam is built to block a river so that the river water forms a lake behind it. The water is then allowed to flow through pipes and turns turbines to generate electricity.

Solar energy

The sun releases enormous amounts of energy. Solar panels can be used to turn some of this energy directly into electricity, or they can be used to heat water for people's homes.

Gasifier

Geothermal heat station

Geothermal energy

Beneath the Earth's crust, there is hot liquid rock. Water heated by this hot rock can be used to heat and cool buildings, or to generate electricity.

Biomass energy

Biomass is any plant or animal material burned as fuel on a large or small scale. It includes wood, plants such as rapeseed, and camel and other animal dung.

Tidal energy

To collect energy from the tides, a long barrier is built at sea. The incoming tide is held behind it, then released. The tide water rushes through turbines in the barrier to produce electricity.

Do I really need this?

We're running out of space for all the garbage we make, and it's damaging our environment. So what can we do? The three Rs are ways of cutting down on waste: reduce—cut down on things you buy; reuse—use things again; recycle—turn old things into new things to save materials, water, and energy.

We're reusing!

Do I need disposable cutlery for my birthday picnic?

No, you don't. Especially if you only use it once, and then throw it away. Use cutlery you can reuse or recycle—the same goes for plates, bowls, and straws.

Do I need wrapping paper?

Not really. Use newspaper instead! If you do buy wrapping paper, choose a type that can be recycled. Some shiny paper is not suitable for this.

Bring your own bag!

Do I need a new plastic bag from the supermarket?

No. Just remember to take your own cloth or reusable plastic bags with you. These can be used many times before they wear out.

Do I need to throw out all my old stuff?

No, there are lots of things you can do with it instead. Why not donate it to a thrift shop for someone else to buy and use, or have a yard sale?

I'm made from a bottle!

Do I need to travel to school in a car?

Not unless you live far away. Walking or cycling safely helps to reduce traffic and air pollution, and keeps you fit and healthy!

Do I need to throw away empty bottles?

The answer is no, once again! Lots of plastic and glass bottles can be recycled, or you can turn them into something artistic or reuse them as containers for storing things.

Do I need to take part in recycling campaigns?

Yes! You'll be helping to clean up the planet, and having fun at the same time. Ask your teacher if you can set up a recycling campaign at school— you'll be amazed at the results! Don't forget to recycle at home, too.

Art made from trash

All over the world, artists and sculptors are turning trash into works of art. Apart from creating unusual and eye-catching pieces, their work also helps to raise awareness about the problems of managing waste, and of reusing and recycling materials.

Gabriel Dishaw, USA

Dishaw admits to digging through dumpsters to find pieces of junk for his projects. He takes apart old typewriters and other machines, then assembles them again.

Haribaabu Naatesan, India

Naatesan creates extraordinary sculptures of cars, giant insects, and other objects from electronic scrap and discarded car parts.

This lion was made from 3,307 lb (1,500 kg) of old car parts.

Belen Hermosa, Spain

Designer Belen Hermosa reused more than 4,000 CDs to build this amazing chair. CDs are made from a plastic that is hard to recycle.

This chair is made from a simple metal frame, covered in rows of CDs.

Sinan Sigic of Atelier Hapax, France

Sigic transforms trash into beautiful objects. These include chess sets made from cardboard and belts made from old gloves.

This bracelet is made from tiny rolls of paper.

These sneakers by Gabriel Dishaw are part of a series of shoes, made from broken computers.

Zero-waste kitchen

In some countries, people struggle to find enough to eat. In others, millions of tons of food are wasted every year. Turning your kitchen into a zero-waste zone means buying less food in the first place, reusing leftovers wherever you can, and throwing less food away.

Oat milk

If you make your own oat milk, you will have lots of leftover oat pulp. Eat the pulp as a porridge or use it to make a cake or muffins.

Oat milk

Feel the bread to see if it has gotten hard.

Stale bread

As long as stale bread hasn't gotten moldy, you can use it to make crispy croutons and breadcrumbs, bread pudding, or French toast. Stale bread can be frozen and used at a later date.

Used coffee grounds

You can sprinkle coffee grounds around your garden plants to help them grow. They also make a nice, smoothing body scrub for your skin. Use it to wash your skin in a warm bath.

Breadcrumbs

Coffee scrub

1 **Buy less** Don't buy too much. Shop little and often so you don't end up throwing lots away.

2 **Be date aware** To avoid food waste, use food before it gets too close to its use-by date.

3 **Use the right containers** Store food properly. This can help it stay fresh and last for longer.

4 **Be tidy** Organize your fridge and cupboards. Know what you need and don't need.

Banana cake

Brown bananas

Overripe bananas might not look good to eat, but don't throw them in the trash. They're delicious made into banana bread or banana ice cream.

Overripe bananas are great in smoothies.

Banana ice cream

Soup

Cheese leftovers

Grate any leftover cheese and add it to your favorite soup, or make a tasty pasta sauce. Use a mixture of different cheeses if you have lots of scraps left.

Tomato sauce

Fruit and veggie peelings

Turn vegetable peel into crunchy chips by spraying them with oil, and cooking them in the oven. Apple peel and cores make a tangy apple tea. You can make a tomato sauce for pasta with overripe tomatoes.

Apple tea

Grated cheese is tasty on toast too.

Decorate a dish with an orange peel flower.

Making soil from scraps

If you've got waste from your kitchen or garden, don't throw it away. Fruit and vegetable peelings, tea bags, dead flowers, and grass clippings can all be used to make a mixture called "compost." The compost can then be spread on flower beds and mixed into soil to help plants to grow.

1. Select a dry and shady place

Put your compost bin on bare, well-drained soil, so that extra water can escape the bin. Choose a warm spot, but out of direct sunlight.

2. Add small layers

Start to fill your bin. Add layers of "green" material (such as weeds, grass, and fruit peelings) and "brown" material (cardboard egg cartons, paper bags, and wool).

3. Keep the compost moist, but not wet

Sprinkle water on each layer as you add them in. You should also water your compost if it starts to dry out. The compost should be kept moist, but not wet.

6. Compost is ready

Keep going until the mixture at the bottom of the bin looks dark and crumbly, like soil. Your compost is ready to use.

5. The pile gets warm

As the mixture rots, your compost heap may start to heat up. You might even see steam rising from the top.

4. Turn the mixture occasionally

Every few weeks, turn the mixture with a shovel. This will let in air, and encourage bacteria to rot down the mixture.

Ways to compost

Compost bins are handy for keeping everything in one place. But if you don't have a bin, you could simply pile your composting scraps in a corner of the garden.

Bins are a neat way to collect scraps.

51

Wash wisely

If your clothes aren't very dirty, wash them on a cold setting in the washing machine. Heating up water for washing uses a lot of energy.

Energy-efficient homes

Most of our homes waste lots of electricity, water, and heat every day. Luckily, there are simple and practical ways of making your home more energy efficient.

Use better bulbs

Change to energy-efficient light bulbs. Not only do they use up to 80 percent less energy than ordinary bulbs, but they also last much longer.

Fix leaky faucets

A leaky faucet in your sink or bath, or a dripping showerhead, can waste thousands of gallons of water each year. Fixing the faucet saves precious water and keeps water bills lower.

It's important to be aware of water waste and to fix leaks as soon as possible.

Eco houses

Forget about bricks! Some eco-friendly houses are built from unusual natural or upcycled materials, such as bales of straw or shipping containers.

A house made from shipping containers in Alaska

A house being made from straw bales

Insulate walls

A large amount of heat is lost through the walls, especially if your house is old. Insulation is a special kind of padding that helps to keep heat in.

Insulation can be made from recycled waste materials.

Turn things off

Make sure you turn off lights, computers, and televisions when you're not using them. And remember that leaving things on standby uses almost as much electricity as when things are switched on.

Mr. Trash Wheel

Trash wheels
Mr. Trash Wheel and Professor Trash Wheel are two funny-looking vehicles that remove trash from the water in Baltimore's Inner Harbor in the United States. The trash is put into dumpsters on board the vehicles.

Vending machine that gives cash for trash

Cash for trash
In Columbia, special vending machines give people rewards for their garbage. When someone deposits a plastic bottle, they get money, movie tickets, or restaurant coupons in return.

Out-of-trash thinking

All that garbage is mounting up, but it's not too late to clean up our act and our planet. New technology for dealing with trash is being developed all the time. Here are some of the interesting ways countries around the world are reducing and recycling waste.

Garbage electricity
Sweden was one of the first countries to take turning trash into electricity seriously. In fact, their technology was so good that they ran out of their own waste, and had to import trash.

Waste that can be turned into electricity

Healthcare exchange

Garbage for medicine
A wonderful program in Indonesia allows people to exchange recyclable trash for medical care. People are advised which items to collect, and where to take them to sell.

Poop to paper

Instead of making paper from wood, why not use animal dung? Elephant, cow, horse, or donkey poop can be turned into eco-friendly, smell-free paper!

Smart bins
Smart bins in Australia can hold five times more trash than a standard garbage can. The bins crush the waste, and sensors sound an alert when the bins are full.

Smart bins

Elephant poop paper products

Meet the expert

Meet Sana Ahmed, a research student at The Open University in the UK. She is researching the dangers of domestic waste. We asked her some questions about her work.

Q: What inspired you as a child?

A: I grew up in Pakistan. At a very young age I had a natural interest in learning new things and solving problems. I have always been fascinated by discoveries and I have a strong desire to help others. I especially enjoyed studying biology (the human body) and microbiology (tiny living things) at school and at university. I have continued developing these interests. That quest for knowledge brought me to the UK, and led me to become a PhD research student.

Q: What are you studying at university?

A: I am researching the dangers that food waste can cause to health. I study the tiny airborne microorganisms that food waste can release. They can be breathed in or ingested through the mouth or nose. My research looks at how this affects the health of people at home using food waste bins, garbage collectors picking up food bags, and waste site workers getting rid of the waste.

Q: What is the best thing about doing your research?

A: No one has ever done this research before, so whatever I discover will be unique. It's also interesting to look at what is happening inside of food waste bins and investigating how it can be a source of health hazards.

Q: What's a typical working day like?

Each day is different, but I mostly spend time in a lab, which is a room with lots of special science equipment. My usual day in the lab includes planning and carrying out experiments, analyzing data, and going to meetings.

Scientists working together in a lab similar to the one that Sana does her experiments in.

Q: What equipment do you use to find out about the dangers of food waste?

A: I use a set of scales to weigh out materials for an experiment. I attach a special sample collector inside a food waste bin (underneath the lid). It gathers tiny airborne particles from food waste in a filter. I then look at these samples under a microscope in a laminar flow hood (a small air-conditioned machine that is very clean and nearly dust-free). I store the samples in a fridge or freezer. I also use an incubator, which is a machine that keeps things at a specific temperature. It allows me to study food waste under a range of temperatures.

Sana sorts through her food waste at home.

Q: Who do you work with?

A: I have colleagues who assist me with my research. I also have teachers who give me feedback and guidance. There are helpful technicians who look after the lab equipment and assist me when I'm working in the lab.

Sana taking out her garbage at home to be collected by a garbage truck

Q: What are your hopes for the future?

A: I believe that my research will aid people who work in the waste industry. Food waste is the third biggest source of greenhouse gas emissions and contributes to global warming. I am hoping that my research can help to provide new ideas on how to treat food waste, which will help to slow down climate change.

Facts and figures

How much garbage do people in New York City throw away every day? How much energy does recycling plastic save? How long does it take for a baby's diaper to rot? Find the answers here as well as other garbage facts.

Save trees!

Making clothes uses lots of water.

Clothing manufacture and sales in the UK is the fourth largest strain on natural resources after housing, transportation, and food.

100,000

marine animals die every year due to plastic waste in seas and oceans.

500 YEARS

is the time it takes for a used disposable diaper to decompose completely.

When **GARDEN WASTE** is dumped in a **LANDFILL**, it releases **METHANE**—a gas that is **23 TIMES MORE HARMFUL** than **CARBON DIOXIDE**.

1,454 ft (443 m) tall

RECYCLING plastic saves twice as much **energy** as burning it.

RECYCLING A **3-FT (1-M)** PILE OF NEWSPAPER **SAVES ONE TREE**.

A trash bag of kitchen waste can **create** enough energy to light a **15 W LED bulb for more** than **six days**.

59 million tons (53 million metric tons) is the average amount of global electronic waste generated each year.

5 MILLION

tons (4.5 million metric tons) of edible food is thrown away in the UK every year.

Glossary

Here are the meanings of some words that are useful for you to know when learning all about garbage.

airborne Something that is transported by the air

archaeologists People who study human history by looking at ancient sites and artifacts

artifacts Objects made by human beings, often long ago

artificial Something made by human beings rather than naturally

atmosphere Thick layer of gases that surround the Earth and protect the planet from the sun's rays

bacteria Microscopic living things, some of which can cause diseases

biodegradable Something that can be rotted down by bacteria or other living things

biofuel A fuel produced from biomass

biomass Materials made from living organisms, such as animals and plants

carbon dioxide A colorless gas which is present in our atmosphere and is absorbed by plants. It is also released by burning fossil fuels. Its symbol is CO_2

centrifuge A machine with a container that spins very quickly so that its contents also spin

climate The weather conditions in a certain area over a long period of time

energy What makes things happen. It is found in different forms, including heat, light, movement, sound, and electricity

experiment A type of scientific test, performed in a laboratory

fertilizer A chemical or natural material added to soil to make crops grow

generator A machine that converts mechanical energy into electricity

global warming An increase in global temperature caused by the greenhouse effect

greenhouse effect The heating up of the Earth as a result of increased greenhouse gas levels, due to human activity

greenhouse gases Gases in the Earth's atmosphere which absorb the sun's heat. Carbon dioxide, methane, nitrous oxide, and water vapor are greenhouse gases

groundwater Water that is found underground in the soil or in rocks

habitat The natural home of an animal or plant

incineration When something is burned to get rid of it

ingested When something is taken into the body through the mouth or nose

media In science, substances in which cells live or are grown

methane A gas with no color or smell, often used as fuel. It is a powerful greenhouse gas that is produced by cattle, burning fossil fuels, or when organic matter decomposes

microbes Tiny living things, such as bacteria

microorganisms Tiny living things, such as bacteria, viruses, or fungi

midden An ancient garbage dump

migrating People or animals who are moving from one place to another

nuclear waste Waste from a nuclear power plant that is dangerous to people, animals, and the environment if it is not disposed of properly

ozone A form of oxygen which creates a layer around the Earth, called the ozone layer, and protects the Earth from the sun's rays. Ozone is pale blue in color

paddy fields Flooded fields where rice is grown

pesticide Chemical sprayed on crops to kill harmful insects and other animals

pulp The soft, fleshy part of a fruit

renewable (energy) Energy produced from a source that is naturally replenished and will not run out, such as wind, solar, tidal, or geothermal energy

sell-by date A date marked on food which shows the date by which it should be sold

single-use Something that is used only once and then thrown away or destroyed

sustainably Done in a way that meets the needs of people today, without damaging the ability of future generations to meet their needs

turbine A machine in which a wheel fitted with blades is made to turn by the flow of liquid or gas. This turning motion produces energy

UV Ultraviolet radiation. A form of energy that travels in the sun's rays

zero-waste Preventing any waste from being sent to landfill or from polluting the environment

WET

DRY

Wet and dry waste

Index

Acknowledgments

Dorling Kindersley would like to thank the following people for their assistance in the preparation of this book: Robin Moul and Dawn Sirett for editorial assistance; Caroline Stamps for proofreading; Helen Peters for compiling the index; and Dan Crisp for illustrations. The publishers would also like to thank Sana Ahmed for the "Meet the expert" interview.

The publisher would like to thank the following for their kind permission to reproduce their photographs:

(Key: a-above; b-below/bottom; c-center; f-far; l-left; r-right; t-top)

1 Haribaabu Naatesan/ www.fossilss.com: (c). 2 Dreamstime.com: Chernetskaya (crb); Zaclurs (bl); Gitanna (fbl, bl/tomatoes); Valiantsina Mironava (bc); Sergeypykhonin (br). 3 123RF.com: Richard Whitcombe (bl). Dreamstime.com: Abrosimovae (tr); Wanuttapong Suwannasilp (br); Winai Tepsuttinun (bc). Tarisio: (cr). 4–5 Dreamstime.com: Panaramka (t). 4 Alamy Stock Photo: Nature Picture Library (bc). 5 Alamy Stock Photo: Tom Vater (clb). Dreamstime.com: GoranJakus (tr). Getty Images: George Osodi / Bloomberg (crb). 6–7 Getty Images: Auscape / Universal Images Group. 6 Dreamstime.com: Mark Eaton (bc); Velichka Miteva (c); Phana Sitti (cb). 7 Alamy Stock Photo: agefotostock (bl); James King-Holmes (cra); Melvyn Longhurst (cla). Dreamstime.com: Claudio Baccaro (c); Sinisa Botas (cl). 8 123RF.com: scanrail (crb). Alamy Stock Photo: PA Images (cr); Hiroko Tanaka (bl). 9 Alamy Stock Photo: Everett Collection Inc (bl). Nick Dimola NYC: (ca, cra). Tarisio: (cr). 10 Alamy Stock Photo: Alistair Scott (cl). Dreamstime.com: Gitanna (cr/Tomatoe, fcr); Sataporn Jiwjalaen / Onairjiw (t); Zaclurs (cr). 10–11 Alamy Stock Photo: Brais Seara (bc). Dreamstime.com: Coppi Bartali (c); Laboko (tc). 11 Dreamstime.com: Damrong Rattanapong (cr). 12 Dreamstime.com: Roman Milert (bl); Photka (bc); Photopips (c); Olena Smyrnova (cra, ca). 12–13 Dreamstime.com: Natalya Danko (bc); Photka (c). 13 123RF.com: bbtreesubmission (cl/Bottles). Dreamstime.com: Ian Allenden (cb); Photka (cl); Khaled Eladawy (tc); Pavel Drozda (c); Hsagencia (clb); Valiantsina Mironava (bc). 14 Alamy Stock Photo: Rubens Alarcon (tl); EnVogue_Photo (cr). 15 123RF.com: eivaisla (cr). Alamy Stock Photo: Arterra Picture Library (tl). Dreamstime.com: Andrey Popov (bl). 16–17 Getty Images: Johner. 17 Dreamstime.com: Al Robinson (tr). Getty Images: Halfpoint Images (tc). 18 Dreamstime.com: Rangizzz (clb); Winai Tepsuttinun (cla); Wanuttapong Suwannasilp (cla/garbage bag). Getty Images / iStock: Mukhina1 (bl). 18–19 Dreamstime.com: Sergeypykhonin (cb); Suphonphoto (bc). Getty Images: Donat Sorokin\TASS (ca). 19 Dreamstime.com: Maldives001 (tr); Romanotino (cr). Getty Images / iStock: Vm (cb). 20 Dreamstime.com: Chernetskaya (cl); Valya82 (cl/watering); Jinaritt Thongruay (cr). Getty Images / iStock: Laughingmango (cr/Vineyard). 21 Dreamstime.com: Weerayuth Kanchanacharoen (cl); Ded Mityay (cl/Water Treatment Plant); Visivasnc (cb/water-saving equipment); Sofiia Shunkina (cr). 23 Dreamstime.com: Chernetskaya (bc); Godruma (clb); Monika Wisniewska (br). 24 Dreamstime.com: Icefront (cr); Marekusz (clb); Vadzim Yakubovich (br, bc). 25 Dreamstime.com: Steve Allen (tr); Comzeal (crb). Getty Images: Pascal Mannaerts / Barcroft Imag / Barcroft Media (cla). Getty Images / iStock: Bet_Noire (bl). 26 NASA: JPL-Caltech (br). 26–27 NASA: European Space Agency. 27 123RF.com: Veniamin Kraskov (cla). Dreamstime.com: Godruma (ca); Lemusique (ca/plastic bag). Getty Images: Stocktrek Images (bc). 28 123RF.com: Olga Yastremska (cl). Dreamstime.com: Olgagillmeister (tr). 29 Dreamstime.com: Olgagillmeister (tc); Golib Tolibov (c); Anton Samsonov (bl). Getty Images / iStock: France68 (bc). 30 Dreamstime.com: Monticelllo (cl). 31 Dreamstime.com: Design56 (cla); Aleksey Popov (crb). Getty Images / iStock: Art_rich (tl); Karayuschij (cla/foam trays); jon11 (cr). 32 123RF.com: Richard Whitcombe (bc). Dreamstime.com: Dmitrii Melnikov (cl). 32–33 Alamy Stock Photo: Paulo Oliveira (c). 33 Getty Images / iStock: doble-d (bl). Science Photo Library: Planetary Visions Ltd (tr). 34–35 Getty Images: Agung Fatma Putra / SOPA Images / LightRocket. 36 Dreamstime.com: Bomboman (br). Getty Images / iStock: Aurore Kervoern (bl). 37 Alamy Stock Photo: Pamela Au (br). Dreamstime.com: Pklimenko (bl); Suthon Thotham (tr). 38 Alamy Stock Photo: Bill Brooks (br). Dreamstime.com: Johnypan (cl). 39 Dreamstime.com: Chingyunsong (cl). Getty Images / iStock: Boryak (tr). Getty Images: Haryadi Bakri / EyeEm (br). 41 Getty Images: Ashley Cooper (tr); Zakir Hossain Chowdhury / Barcroft Media (tc). 42 Dreamstime.com: Ultramarine5 (crb). Getty Images / iStock: Pifate (bl). 43 Alamy Stock Photo: Universal Images Group North America LLC / DeAgostini (tl). Dreamstime.com: Péter Gudella (tl); Nostal6ie (tr). 44 Dreamstime.com: Chernetskaya (crb); Wavebreakmedia Ltd (cl); Olga1969 (clb). Getty Images: Photos by Sally Jane Photographic Art (fclb). 45 Dreamstime.com: Abrosimovae (cl); Motortion (tl); Monkey Business Images Ltd (cra); Wavebreakmedia Ltd (br). 46–47 Alamy Stock Photo: WENN Rights Ltd / Gabriel Dishaw (c). 46 Haribaabu Naatesan/ www.fossilss.com: (b). 47 Belen Hermosa: (tr). Sinan Sigic: (br). 48 Dreamstime.com: Chernetskaya (cb, br); Volodymyr Muliar (cl); Photomailbox (bl); Oleksandra Naumenko (cra). 49 Dreamstime.com: Nina Firsova (cra); Anouk Stricher (tr); Sergii Petruk (ca); Hans Geel (cl); Andrey Starostin (c); Tashka2000 (clb); Zulfiia Ishmukhametova (bc); Aleksandr Volkov (bc). 51 Shutterstock.com: runzelkorn (br). 52 Getty Images: Peter Dazeley (tl); Jose Luis Pelaez Inc (bc). 52–53 123RF.com: Andriy Popov (tc). 53 Alamy Stock Photo: Jeff Morgan 15 (tr). Dreamstime.com: Andrii Biletskyi (cr); Rpianoshow (tc); Manaemedia (br). 54–55 Dorling Kindersley: Merritt Cartographic: Ed Merritt (c). 54 TOMRA: (cl). Waterfront Partnership of Baltimore: (tl). 55 Alamy Stock Photo: hemis.fr / Gil Giuglio (br); P&F Photography (bl). Dreamstime.com: Mykola Sirenko (tr); Wanuttapong Suwannasilp (b). Getty Images: Aman Rochman / AFP (c). 56 Dreamstime.com: Mengtianhan (bl). Ahmed Nawaz: (tl). 57 Ahmed Nawaz: (bl, tr). 58–59 Alamy Stock Photo: Thesimplegraphy (tc). 58 Dreamstime.com: Blueringmedia (bl); Jemastock (cl); Risto Hunt (crb); Onyxprj (br). 59 Dreamstime.com: Frions (br); Konstantin Gorbachev (r); Laboko (cb); Damrong Rattanapong (bl). 62 Alamy Stock Photo: Paulo Oliveira (tl). 64 Dreamstime.com: Zulfiia Ishmukhametova (ftl); Aleksandr Volkov (tl)

Endpaper images: Front: Alamy Stock Photo: Ian Dagnall cb, Richard Levine fcrb, Niday Picture Library ca (White Wings), Prisma Archivo fbl; Dreamstime.com: Xiaoma ftl; Getty Images: Andrew Holbrooke crb, Hulton Archive tc; Library Company of Philadelphia: bc (Roberts' Old Mill); Science Photo Library: bc; Back: 123RF.com: yasonya cra; Depositphotos Inc: NosorogUA clb; Dreamstime.com: Kasia Biel cr, Nadezhda Bugaeva cb, Caymia bl, Chernetskaya br, Duskbabe bc, Elvira Koneva c, Pamela Mcadams bc (Cloth Diapers), Anatoly Repin cla

Cover images: Front: Dreamstime.com: Jemastock crb, Lemusique cra, Olga1969 ca, Pixelrobot bl, Stockernumber2 c; Getty Images: Photos by Sally Jane Photographic Art ca/ (Gift); Getty Images / iStock: Bet_Noire cra/ (barrel); Back: Dreamstime.com: Zulfiia Ishmukhametova tr, Aleksey Popov tl; Spine: Dreamstime.com: Frions b.

All other images © Dorling Kindersley
For further information see: www.dkimages.com

My Findout facts:

Dispose or reuse?

Liquid soaps

Liquid soap comes in plastic bottles that use lots of energy to make. Also, the empty bottles do not break down, when you throw them away.

Single-use nappies

Single-use nappies are made from plastics which are not biodegradable. They are dumped in landfill, where they can take hundreds of years to break down.

Toothpaste tubes

Millions of toothpaste tubes are thrown away every year. They are difficult to recycle because they are made of a mix of plastic and other materials.

Toothpaste

Extra White

Plastic packaging

Plastic packaging helps to keep food fresh, so that it can be transported over long distances. But much of it ends up in landfill, where it may never break down.

Plastic storage

Some plastic food containers leak harmful chemicals into food, when they are heated or microwaved. Many are made of plastic that cannot be recycled.